My Last ABC Book

Brad Collins

ISBN: 978-1-7645521-0-3

This book is dedicated to Stella
aka Puss Puss, Catrix, Long Cat, Rabbit, Possum.

A

B

C

D

E

F

G

H

Caturday
49 20 43 41 4E
20 48 41 53 20
43 48 45 45 5A
42 55 52 47 45
52 3F

I

J

K

L

M

N

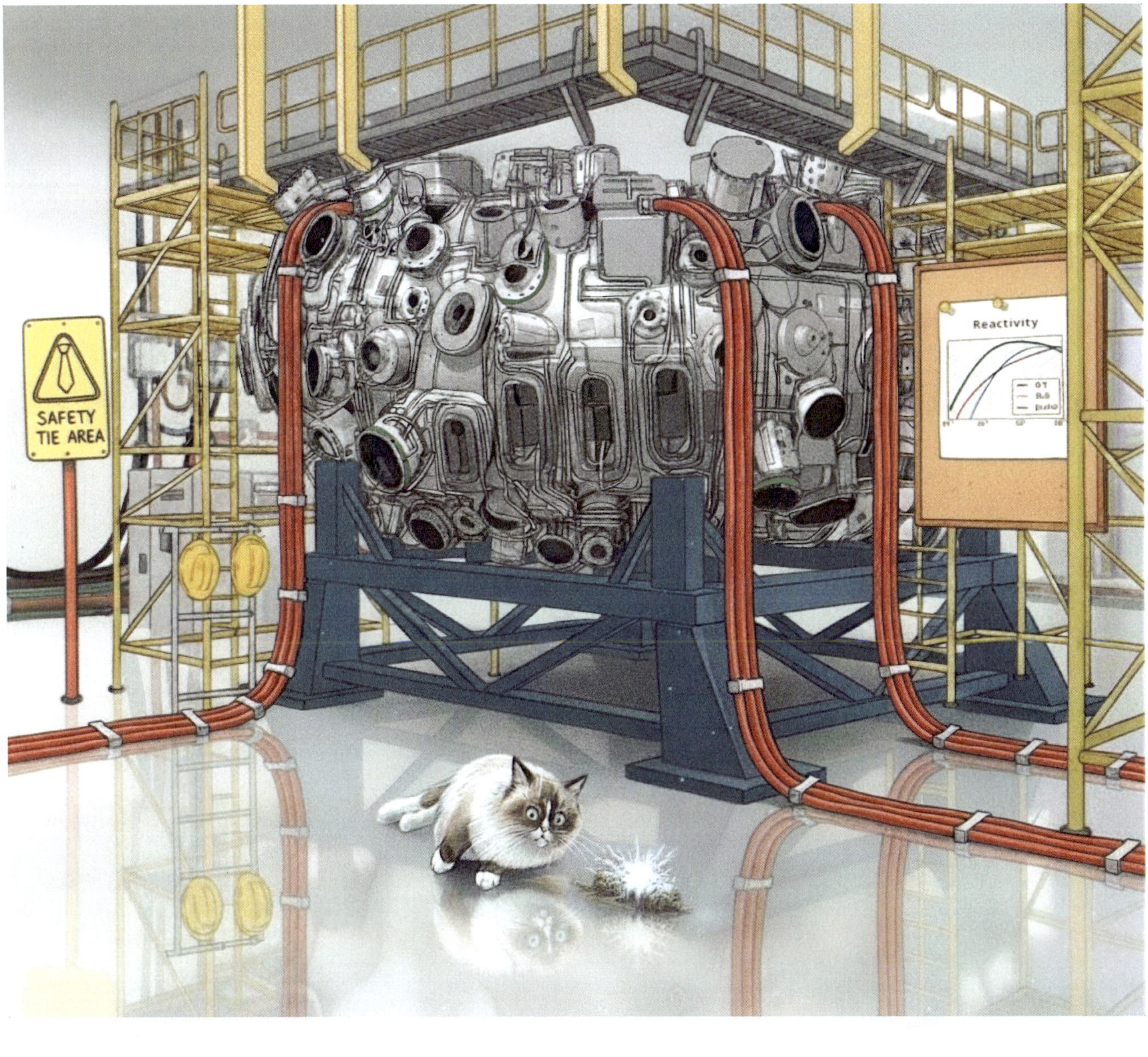

O

P

Q

R

$C = M^e \bmod n$
$\Phi(n) = \Phi(p * q) = \Phi(p) * \Phi(q) = (p - 1)*(q - 1)$
$(d * e) \equiv 1 \bmod \Phi(n)$
Sgvello
oorldd
sgT5y
h85y h8F1m
BIG
PRIMES

S

T

U

V

W

X

Y

Z

Answers:

A is for Apple, although Stella obviously prefers Arch Linux.

B is for Ball. Stella usually prefers playing with a piece of string. So much so, she's practically mastered string theory.

C is for Chef. Stella loves to help in the kitchen, especially when the steak or mince comes out.

D is for Dinosaur. Stella's favourite dinosaur is the Australian Magpie that visits occasionally.

E is for Electricity. Don't let the magic smoke out.

F is for Fibonacci sequence. A series of numbers where each number is the sum of the two preceding numbers.

G is for Gnomon. The part of a sundial that casts the shadow and indicates the time (Standard Time rather than Solar Time in the case of this Richard Schmoyer sundial with its clever analemma design).

H is for Hexadecimal, a base-16 positional numeral system often used in computing for colour codes, memory addresses, MAC addresses and ASCII character encoding.

I is for Imaginary numbers, the numbers Stella chases when she has the zoomies.

J is for Jackpot. Stella will win big if you buy this book for all your family and friends.

K is for Knot theory. The study of closed curves in three dimensions. Stella is confused by the recent disproval of the Unknotting Number Additivity Conjecture.

L is for L2 Lagrangian point. Home of the James Webb Space Telescope, and the location where the combined gravitational forces of the Sun and Earth provide the exact centripetal force needed for an object to maintain a fixed relative position to both bodies (think restricted three-body problem).

M is for Mandelbrot set. A famous fractal exhibiting self-similarity; meaning its intricate pattern repeats infinitely at every level of magnification.

N is for Nuclear fusion. The process where two light atomic nuclei combine to form a single, heavier nucleus, releasing a vast amount of energy in the process. Stella believes the Stellarator reactor design is promising, and thinks it should only be 10 years until $Q_E > 1$.

O is for Ornithopter. An aircraft that flies by flapping its wings.

P is for Portal. Stella gives her new cat flap a five o-type star rating.

Q is for Quasar: The extremely bright core of an ancient galaxy, powered by a supermassive black hole consuming vast amounts of matter. Not to be confused with the Cat Head Nebula.

R is for RSA encryption. If $C = 2557$, $n = 3233$, and $d = 413$, then $M = ?$

S is for Spacetime. Stella uses her mass to bend the fabric of spacetime, guaranteeing herself a longer, gravitationally-dilated sleep.

T is for Telekinesis. The supposed ability to move or manipulate objects using only the power of the mind.

U is for Unsaturated hydrocarbon. An organic compound composed only of carbon and hydrogen atoms and containing at least one carbon-carbon double bond or carbon-carbon triple bond.

V is for Vacuum tube. Stella is fascinated by vacuum tubes, which is weird, because she hates vacuum cleaners.

W is for Wolfram, aka tungsten. It's not actually possible to melt wolfram in a typical forge due to its incredibly high melting point.

X is for Xenon. A noble gas with many interesting uses—from ion-thrusters to anaesthesia.

Y is for Yagi-Uda antenna. A highly directional antenna design often used for long range communications like UHF/VHF radio. Build your own using a tape measure.

Z is for Zeppelin. The LZ 129 Hindenburg zeppelin was the largest aircraft ever built (by length and volume).

www.ingramcontent.com/pod-product-compliance
Lightning Source LLC
LaVergne TN
LVRC091049081126
851858LV00030B/291